Renewable Energy Resources

Dr. Hemant Pathak

DEDICATION

Dedicated to Shri Sainath Maharaj the all omnipotent of world the most merciful.

CONTENTS

Foreword

Energy supply from renewable resources is now essential global strategy, especially when there is responsibility for the environment and for sustainability.

The world is moving towards a sustainable energy future with an emphasis on energy efficiency and use of renewable energy sources. A finite planet cannot support infinitely increasing consumption of resources and hence the motto of present times must be to 3 R principal - "Reduce, Reuse, Recycle".

Renewable Energy Resources; provides a unique insight into the problems our planet faces in terms of clean energy resources proven technical and economic importance worldwide. This books Written for academics, researchers and practitioners working in Energy field, expressed comprehensive and interdisciplinary focus on the current energy demand to enhance energy conservation outcomes.

This book has been provided to be utilized by all people concerned with energy conservation in all the industries in world.

This book provides an essential guide to researchers, it offers: various aspects of Renewable energy resources; on the challenges and experiences in present scenario.

Simply explained, Renewable Energy Resources is an important book bringing together diverse viewpoints from Industries and state agencies and regulators, for all who wish to make a difference in how to plan and manage our Energy resources.

Dr. Hemant Pathak

M.Sc. (Gold medalist), Ph. D.

Assistant Professor of Engineering Chemistry

Indira Gandhi Govt. Engineering college, Sagar, MP, India

Acronymns

CH_4 methane

CNG compressed natural gas

CO_2 carbon dioxide

CO_2e CO2 equivalent

CPP Critical Peak Pricing

CRIS Climate Registry Information System

ECMP Energy Conservation and Management Plan

EMAP Energy Management Action Plan

EMG Energy Management Group

EMS Environmental Management System

GHG greenhouse gas

GIS Geographic Information System

N2O nitrous oxide

NPV net-present value

O&M operations and maintenance

RECs renewable energy credits

REP Renewable Energy Program (Policy)

Energy units and conversion factors

Temperature Kelvin (K)

Commonly used temperature units

Celsius (C), Fahrenheit (F)

$0°C = 273.15 \text{ K} = 32°F \ 1°F = 5/9°C \ 1°C = 1 \text{ K}$

Fahrenheit temperature = 1.8 (Celsius temperature) + 32

Derived SI units

Heat: Quantity of heat, work, energy joule (J)

Heat flow rate, power watt (W)

Heat flow rate watt/m2

Thermal conductivity W/mK

Glossary

Abatement   The reduction or elimination of pollution.

Acid rain   The precipitation of dilute solutions of strong mineral acids, formed by the mixing in the atmosphere of various industrial pollutants

Act   A law

Aerosol   Particles of solid or liquid matter than can remain suspended in air from a few minutes to many months depending on the particle size and weight.

Air pollution   Toxic or radioactive gases   or particulate matter introduced into the atmosphere, usually as a result of human activity.

Air Toxic   Any air pollutant for which a ambient air quality standard does not exist that may reasonably be anticipated to cause cancer, developmental effects, reproductive dysfunctions, neurological disorders, heritable gene mutations or other serious or irreversible chronic or acute health effects in humans.

Ash   Incombustible residue left over after incineration or other thermal processes.

Bagasse   The fiber residue that remains after juice extraction from sugarcane.

Btu   The abbreviation for British Thermal Unit(s).

Byproduct   A secondary or additional product resulting from the feedstock use of energy or the processing of nonenergy materials. For example, the more common byproducts of coke ovens are coal gas, tar, and a mixture of benzene, toluene, and xylenes (BTX).

Bioenergy   The conversion of biomass into useful forms of energy such as heat, electricity and liquid fuels.

Biogas   Gas produced by the biological process of anaerobic (without air) digestion of organic material.

Biomass   Organic, non-fossil material of biological origin constituting an exploitable energy source.

Carbon dioxide (CO_2)   A colorless, odorless, non-poisonous gas that is a normal part of Earth's atmosphere. Carbon dioxide is a product of fossil-fuel combustion as well as other processes. It is considered a greenhouse gas as it traps heat (infrared energy) radiated by the Earth into the atmosphere and thereby contributes to the potential for global warming. The global warming potential (GWP) of other greenhouse gases is measured in relation to that of carbon dioxide, which by international scientific convention is assigned a value of one .

Carbon sink   A reservoir that absorbs or takes up released carbon from another part of the carbon cycle. The four sinks, which are regions of the Earth within which carbon behaves in a systematic manner, are the atmosphere, terrestrial biosphere (usually including freshwater systems), oceans, and sediments (including fossil fuels)

Climate change   A regional change in temperature and weather patterns. Current science indicates a discernible link between climate change over the last century and human activity, specifically the burning of fossil fuels.

Clean   Energy   Electricity and/or heat producing systems that
Technologies   produce negligible or minimal amounts of environmental pollution compared with conventional technologies.

Coal   A readily combustible black or brownish-black rock whose composition, including inherent moisture, consists of more than 50 percent by weight and more than 70 percent by volume of carbonaceous material. It is formed from plant remains that have been compacted, hardened, chemically altered, and metamorphosed by heat and pressure over geologic time.

Combustion   Burning. Many important pollutants, such as sulfur dioxide, nitrogen oxides, and particulates (PM-10) are combustion products, often products of the burning of fuels such as coal, oil, gas, and wood.

Contamination   The act of polluting or making impure; any indication of chemical, sediment, or biological impurities.

Dust   Solid particulate matter that can become airborne.

Ecosystem   An interactive system that includes the organisms of a natural community association together with their abiotic physical, chemical, and geochemical environment.

Electric current   The flow of electric charge. The preferred unit of measure is the ampere.

Electric energy   The ability of an electric current to produce work, heat, light, or other forms of energy. It is measured in kilowatthours.

Electric generation industry   The electric generation industry includes the "electric power sector" (utility generators and independent power producers) and industrial and commercial power generators, including combined-heat-and-power producers, but excludes units at single-family dwellings.

Emission   Release of pollutants into the air from a source. We say sources emit pollutants. Continuous emission monitoring systems (CEMS) are machines, which some large sources are required to install, to make continuous measurements of pollutant release.

Emissions coefficient   A unique value for scaling emissions to activity data in terms of a standard rate of emissions per unit of activity

Electricity   A form of energy characterized by the presence and motion of elementary charged particles generated by friction, induction, or chemical change.

Energy   The capacity for doing work as measured by the capability of doing work (potential energy) or the conversion of this capability to motion (kinetic energy).

Energy consumption   The use of energy as a source of heat or power or as a raw material input to a manufacturing process.

Energy Audit   An assessment of a home's energy use. These include a number of different types of surveys, including (in increasing order of cost and complexity): online audits, in-home home energy surveys, diagnostic home energy surveys, and comprehensive home energy audits.

Energy Conservation   Saving energy by doing with less or doing without (e.g., setting thermostats lower in winter and higher in summer; turning off lights; taking shorter showers; turning off air conditioners; etc.).

Energy Efficiency   A ratio of service provided to energy input. Services provided can include buildings-sector end uses such as lighting, refrigeration, and heating: industrial processes; or vehicle transportation. Unlike conservation, which involves some reduction of service, energy efficiency provides energy reductions without sacrifice of service.

Energy loss   Deleted because there is no need for a general term to encompass all forms of energy loss. Terms referring to losses specific to particular energy sources are defined separately.

Exposure   The concentration of the pollutant in the air multiplied by the population exposed to that concentration over a specified time period.

Fossil fuels   Fuels such as coal, oil, and natural gas; so-called because they are the remains of ancient plant and animal life.

Gasification   Combustible gas called producer-gas produced from biomass through a high temperature thermo-chemical process. Involves burning biomass without sufficient air for full combustion, but with enough air to convert the solid biomass into a gaseous fuel.

Geothermal   Natural heat extracted from the earth's crust using its vertical thermal gradient, where discontinuity in the earth's crust

Global warming   increase in the average temperature of the earth's surface.

Greenhouse gases   Atmospheric gases such as carbon dioxide, methane, chlorofluorocarbons, nitrous oxide, ozone, and water vapor that slow the passage of re-radiated heat through the Earth's atmosphere.

Hydrocarbons   An international agreement adopted in December 1997 in Kyoto, Japan. The Protocol sets binding emission targets for developed countries that would reduce the emissions on average 5.2 percent below 1990 levels.

Industrialized countries   The metric prefix for one millionth of the unit that follows.

Kyoto Protocol   An international treaty created in 1997 in Kyoto, Japan to reduce industrial nation's global emissions of greenhouse gases.

Megawatts   one million watts or one thousand kilowatts

Methane (CH$_4$)   A gas emitted from coal seams, natural wetlands, rice paddies, enteric fermentation (gases emitted by ruminant animals), biomass burning, anaerobic decay or organic wastes in landfill sites, gas drilling and the activities of termites.

Photovoltaics   The use of lenses or mirrors to concentrate direct solar radiation onto small areas of solar cells, or the use of flat-plate photovoltaic modules using large arrays of solar cells to convert the sun's radiation into electricity.

Renewable Energy Resources

1. Introduction

In 2002, Johannesburg, South Africa for the 10th anniversary of the first Earth Summit in Rio de Janeiro, Brazil. At the Johannesburg Earth Summit, countries committed to:

" **Substantially increase the global share of renewable energy sources with the objective of increasing its contribution to total energy supply"**

At the 20th Earth Summit June 2012, countries, companies, cities, and individuals need to commit to increasing the amount of wind, solar, geothermal, tidal, and wave power throughout the world to 15 percent of total electricity by 2020 more than doubling what is predicted under current trends. Civic and corporate stakeholders must commit to do more to increase electricity production from renewable sources.

Energy is the driver of growth. Energy is one of the most important building block in human development, and essential factor in determining the economic development of any nation. Energy, irrespective of its form is a scarce commodity and a most valuable resource. The potential of renewable energy sources is enormous as they can in principle meet many times. Energy is one of the most important resources to sustain our lives. At present we still depend a lot on fossil fuels and other kinds of non-renewable energy.

In general renewable forms of energy are considered "green" because they cause little depletion of the Earth's resources, have beneficial environmental impacts, and cause negligible emissions during power generation.

Non-renewable sources include fossil fuels, e.g. oil, coal, gas and their deposits are limited and can be exhausted. Renewable energy sources include solar, wind, biomass, hydro, geothermal and ocean power.

Non-renewable energy sources are responsible for the greenhouse effect, causing global warming, which endangers our planet and future generations. The world's energy demand. Renewable energy sources such as biomass, wind, solar, hydropower, and geothermal can provide sustainable energy services, based on the use of routinely available, indigenous resources. A transition to renewable based energy systems is looking increasingly likely as their costs decline while the price of oil and gas continue to fluctuate.

Renewable energy supplies are of ever increasing environmental and economic importance in all countries. A wide range of renewable energy technologies are established commercially and recognized as growth industries.

International studies on human development indicate that India needs much larger per capita energy consumption to provide better living conditions to its citizens.

India's second largest population and increasing pace of economic growth make its energy needs particularly challenging.

India is now the eleventh largest economy in the world, fourth in terms of purchasing power. It is poised to make tremendous economic strides over the next ten years, with significant development already in the planning stages. This report gives an overview of the renewable energies market in India.

India is an emerging economy, increasing GDP is driving the demand for additional electrical energy, as well as transportation fuels.

Fossil fuel and renewable energy prices, and social and environmental costs are heading in opposite directions and the economic and policy mechanisms needed to support the widespread dissemination and sustainable markets for renewable energy systems are rapidly evolving.

India has a vast supply of renewable energy resources, and it has one of the largest programs in the world for deploying renewable energy products and systems. The extensive use of renewable energy including solar energy needs more time for technology development.

In G20 countries, Germany had the largest amount of its electricity produced from renewable sources in 2011, followed by the European Union, Italy and Indonesia. The United States ranked 7th, India came in 9th, and China ranked 12th. Spain, Portugal, Iceland, and New Zealand which each produced more than 15 percent of their electricity from these sources.

In India 3,700 MW are currently powered by renewable energy sources. The key drivers for renewable energy are the following:

o Due to population increases filling of demand-supply gap

o Eco- friendly, Enormous potential

o Necessity to strengthen India's energy security

o Pressure on high-emission industry sectors from their shareholders

o For rural electrification.

India's energy requirement comes mainly from five sectors; industry, agriculture, transport, services and domestic, each having considerable saving potential.

Energy needs of the country, forecasts of consumption and production, and we assess whether India can power its growth and its society with renewable resources.

Human population and the individual life expectation will increase, energy could, in the future, be in short supply. Unless that supply is increased, it will be a source of friction in human affairs energy conservation is the deliberate practice or an attempt to save electricity, fuel oil or gas or any other combustible material, to be able to put to additional use for additional productivity without spending any additional resources or money.

The development and use of renewable energy sources can enhance diversity in energy supply markets, contribute to securing long term sustainable energy supplies, help reduce local and global atmospheric emissions, and provide commercially attractive options to meet specific energy service needs, particularly in developing countries and rural areas helping to create new employment opportunities there.

To better understand the current situation in India and the future of the renewable energies market, it is important to look at the trends in energy consumption, growth of the current grid, and the availability of transportation and equipment used there.

The development and deployment of renewable energy, products, and services in India is driven by the need to □decrease dependence on energy imports sustain accelerated deployment of renewable energy system and devices expand cost effective energy supply □

These fossil fuel based energy sources are facing increasing pressure on a host of environmental fronts, with perhaps the most serious challenge confronting the future use of coal being the Kyoto Protocol greenhouse gas (GHG) reduction targets.

It is now clear that any effort to maintain atmospheric levels of CO_2 below even 550 ppm cannot be based fundamentally on an oil and coal-powered global economy, barring radical carbon sequestration efforts.

2. Solar Energy

The sun is the most powerful source of energy and this energy is free. Energy provided by the sun through radiation. Technologies are categorized as either active or passive.

Active technologies convert solar energy into a electrical or thermal. Photovoltaic cells that convert sunlight directly into electrical energy, the solar collectors for domestic hot water heating or even solar space heating and cooling, the solar concentrators that use mirrors to focus solar irradiation and generate intense heat, turning water to steam and generating electricity using certain machines and even solar ovens.

Passive technologies seek to place buildings in a favorable orientation towards the sun and architectural designs to exploit solar energy.

The potential of renewable energy sources is enormous as they can in principle meet many times the world's energy demand. Renewable energy sources such as biomass, wind, solar, hydropower, and geothermal can provide sustainable energy services, based on the use of routinely available, indigenous resources.

Solar represents a small share of the electric market in the United States about ½ of one percent of electrical capacity. Solar's contribution to heating and lighting is much larger.

3. **Wind Energy**

The energy of wind used either in windmills or in sailing. Wind energy produce by specific blades to capture wind and machines to transform it to electrical energy. Wind turbines are installed both onshore and offshore in places where wind speed is generally high and constant. It is environment friendly, clean and safe energy resources. ☐It has the lowest gestation period as compared to conventional energy and maintenance costs are low. There is no fuel consumption, hence low operating costs.☐The capital cost is lowest comparable with conventional thermal power plants. For a wind farm, the capital cost ranges between Rs. 4.5

crores to 5.5 crores, depending on the site and the wind electric generator selected for installation.

4. Biomass

Biomass has been used since man invented fire and used to burn wood to heat or cook. Biomass includes solid biomass (organic, non-fossil material of biological origins), biogas (principally methane and carbon dioxide), liquid bio-fuels, and municipal waste.

Biomass is second to hydropower as a leader in renewable energy production. Biomass has an existing capacity of over 7,000 MW. Biomass as a fuel consists of organic matter such as industrial waste, agricultural waste, wood, and bark.

Biomass can be burned directly in specially designed power plants, or used to replace up to15% of coal as a fuel in ordinary power plants. Biomass burns cleaner than coal because it has less sulfur, which means less sulfur dioxide will be emitted into the atmosphere.

Biomass can also be used indirectly, since it produces methane gas as it decays or through a modern process called gasification.

Energy embodied in plants and organic material. Plant biomass comes from the sun through the photosynthesis process, when they capture solar energy.

Biomass includes a wide variety of materials including wood, energy crops, agricultural and forest residues, food waste and organic components from municipal and industrial waste.

The important forms of biomass are sugar cane bagasse in agriculture, pulp and paper residues in forestry and manure in livestock residues. It is argued that biomass can directly substitute fossil fuels, as more effective in decreasing atmospheric CO_2 than carbon sequestration in trees.

A number of conversion technologies exist to convert biomass energy into other form. These technologies converts the energy in forms that can be used directly viz. heat or electricity also liquid bio fuel or combustible biogas.

The Kyoto Protocol encourages uses of biomass energy. Biomass may be used in a number of ways to produce energy. Important methods are:

- Gasification
- Combustion
- Fermentation
- Anaerobic digestion

India has 537 MW commissioned and 536 MW under construction. India is very rich in biomass. It has a potential of 19,500 MW (3,500 MW from bagasse based cogeneration and 16,000 MW from surplus biomass).

Following is a list of some States with most potential for biomass production:

- Andhra Pradesh (200 MW)
- Bihar (200 MW)
- Gujarat (200 MW)
- Karnataka (300 MW)
- Maharashtra (1,000 MW)
- Punjab (150 MW)
- Tamil Nadu (350 MW)
- Uttar Pradesh (1,000 MW)

The largest use of biomass energy in USA is the forest products industry. Furniture plants, sawmills, and paper mills usually burn their wood waste to produce heat and electricity.

5. Hydropower

Water is an excellent and abundant natural resource for generating renewable energy. Water also, is a renewable energy source since it is recharged through the cycle of evaporation and precipitation. Its power was known since ancient years and was exploited through dams, water mills and irrigation systems. The energy of the falling or moving water can be harnessed by various technologies. Water wheels can transform it directly into mechanical energy, turbines and electrical generators can transform it into electricity.

Hydropower is the largest renewable resource used for electricity. It plays an essential role in many regions of the world with more than 150 countries generating hydroelectric power.

Hydropower is a significant source of electricity worldwide and will likely continue to grow especially in the developing countries. Hydropower represents one of the oldest and largest renewable power sources and accounts for close to 10% of Indian electricity. Existing hydropower capacity is about 80,000 megawatts. Hydropower plants convert the energy of flowing water into electricity. Hydropower continues to be the most efficient way to generate electricity.

Modern hydro turbines can convert as much as 90 percent of the available energy into electricity. The best fossil fuel plants are only about 50 percent efficient.

A survey in 1997 by The International Journal on Hydropower & Dams found that hydro supplies at least 50 percent of national electricity production in 63 countries and at least 90 percent in 23 countries. About 10 countries obtain essentially all their commercial electricity from hydro, including Norway, several African nations, Bhutan and Paraguay.

6. Geothermal Energy

Geothermal energy stored and created inside the earth in the form of thermal energy. This energy is released to the surface through volcanoes, geysers hot springs. This can be achieved from deep underground reservoirs through drilling, or from other geothermal reservoirs closer to the surface.

The amount of geothermal energy is enormous. Scientists estimate that just 1 percent of the heat contained in just the uppermost 10 kilometers of the earth's crust is equivalent to 500 times the energy contained in all of the earth's oil and gas resources.

Geothermal energy can be used in residential applications also, e.g. small geothermal heat pumps.

Geothermal electric capacity in the USA is approx. 3,500 MW. Geothermal power plants use high temperatures deep underground to produce steam, which then powers turbines that produce electricity. Geothermal power plants can draw from underground reservoirs of hot water or can heat water by pumping it into hot, dry rock. Geothermal energy can be harnessed to produce electricity or for heating and cooling purposes.

In Geothermal exploitation process variable concentrations of gases, nitrogen and carbon dioxide with some hydrogen sulphide and smaller proportions of ammonia, mercury, radon and boron. Most of these chemicals are concentrated in the disposal water which is usually reinjected back into the drill holes so that there is minimal release into the environment.

The concentrations of the gases are usually low enough not to be harmful or else the abatement of toxic gases can be managed with modern technology.

7. Ocean or Marine Energy

Mechanical energy carried by ocean waves and tides or to the thermal energy of the ocean coming from the sun. Covering almost 70% of earth surface, oceans may prove to be the renewable energy of the future, however harnessing ocean energy to produce electricity is not cost-effective currently.

8 . References

1. Energy for a sustainable world: Jose Goldenberg, Thomas Johansson, A.K.N.Reddy, Robert Williams (Wiley Eastern).

2. Modeling approach to long term demand and energy implication : J.K.Parikh.

3. Energy Policy and Planning : B.Bukhootsow.

4. World Energy Resources : Charles E. Brown, Springer2002.

5. 'International Energy Outlook' -EIA annual Publication

6. Principles of Energy Conversion: A.W. Culp (McGraw Hill International edition.)

7. Aspects of Energy Conversion : I.M.Blair and B.O.Jones

8. Principles of Energy Conversion : A.W.Culp (McGrawHill International)

9. Energy conversion principles : Begamudre , Rakoshdas

10. Principles of Energy Conversion : A.W. Culp.

11. Energy Management: W.R.Murphy, G.Mckay (Butterworths).

12. Energy Management Principles: C.B.Smith (Pergamon Press).

13. Efficient Use of Energy : I.G.C.Dryden (Butterworth Scientific)

14. Energy Economics -A.V.Desai (Wieley Eastern)

15. Industrial Energy Conservation : D.A. Reay (Pergammon Press)

16. Energy Management Handbook – W.C. Turner (John Wiley and Sons, A Wiley Interscience Publication)

17. Efficient Use of Energy: I.G.C.Dryden (Butterworth Scientific)

18. Energy Management Handbook – W.C. Turner (John Wiley and Sons, A Wiley Interscience publication)

19. Industrial Energy Management and Utilisation –L.C. Witte, P.S. Schmidt, D.R. Brown (Hemisphere Publication, Washington, 1988)

20. http://learn-energy.net/education/renewables.php

21. The European Geothermal Energy Council (EGEC), http://egec.info/publications/

22. http://www.eu-oea.com/technology-2/

23. http://energyquest.ca.gov/story/index.html

ABOUT THE AUTHOR

Dr. Hemant Pathak held positions as Assistant Professor in the department of chemistry, Govt. Indira Gandhi Engineering College, Sagar, MP, India. He had extensive experience in teaching, research and administrative management.

Dr. Pathak received his Ph.D. degree in chemistry from Dr. Hari Singh Gour Central University, Sagar, India and M.Sc. Gold medalist from Jiwaji University, Gwalior. He has published 18 books and more than 50 research papers in reputed International and National journals and received several awards. He is a member of editorial boards and reviewer boards of several international journals and societies. His area of specialization includes Engineering Chemistry, Energy audits and Environmental Pollution management.